Department for Transport, Local Government and the Regio

AN INTRODUCTION TO THE USE OF VEHICLE ACTUATED PORTABLE TRAFFIC SIGNALS

London: The Stationery Office

Twelfth Impression 2002

ISBN 0 11 550781 7

Following the General Election in June 2001 the responsibilities of the former Department of the Environment, Transport and the Regions (DETR) in this area were transferred to the new Department for Transport, Local Government and the Regions (DTLR).

SIGNAL CONTROL

should always be

VEHICLE ACTUATED (VA)

(except where otherwise instructed in writing by the Highway Authority)

You will find that modes other than VA are provided on the controller but these should only be used to relieve short term difficulties.

these are:

MANUAL (MAN) —**should be used to stop traffic if the shuttle lane has to be occupied for short periods (e.g. for unloading).**

FIXED TIME (FT) —**may be used while awaiting the arrival of the service engineer if the equipment needs attention.**

THE OBJECT OF USING VEHICLE ACTUATION

To reduce delay to traffic by ensuring that the time for which the green signal is shown is automatically adjusted to suit the amount of traffic using the road.

WHAT IS VEHICLE ACTUATION?

Modern portable traffic signals are operated automatically by the approaching vehicles.

If there is no approaching traffic the signals stay at red in both directions.

The first vehicle to arrive will cause the lights to change to green—allowing it to pass

Following traffic will ensure that the green light is maintained — the signals will not normally change back to red until the last vehicle has passed, but if there is no traffic waiting at the other end the green light may continue to be shown.

THE LENGTH OF TIME FOR WHICH THE GREEN LIGHT WILL SHOW WILL VARY ACCORDING TO TRAFFIC DEMAND.

HOW DOES VEHICLE ACTUATION WORK?:

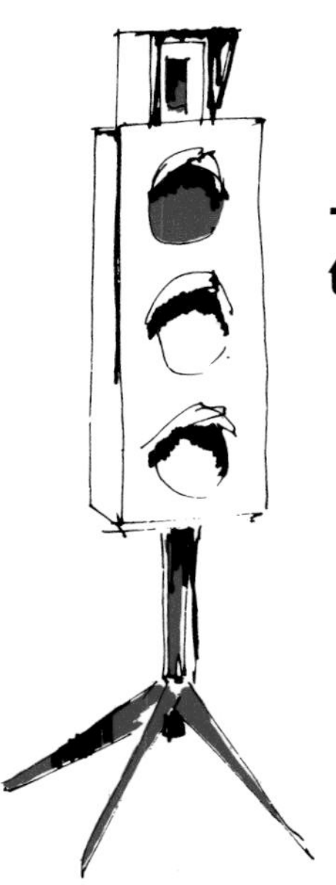

Each signal is provided with a small "radar" unit, often referred to as an 'MVD' (MICROWAVE VEHICLE DETECTOR)

—some are fixed to the top of the signal head—

—others have to be fitted on site to a bracket on the tripod.

The MVD can detect moving vehicles up to 70 metres away — providing they are travelling towards the MVD at speeds greater than 5 mph.

The MVD sends signals to the signal controller along the connecting cables—

TO BE ABLE TO USE THIS METHOD THE CONTROLLER MUST BE SWITCHED TO 'VA'.

REMEMBER!

MVD'S CANNOT WORK PROPERLY IF:

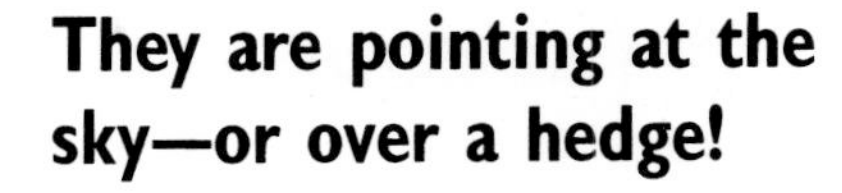

They are pointing at the sky—or over a hedge!

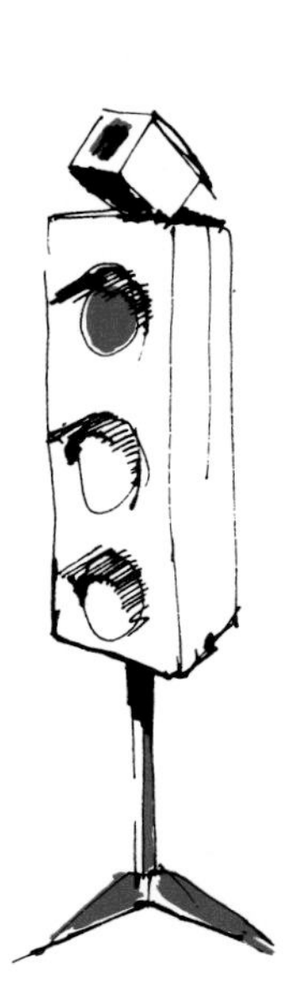

They do not face the oncoming traffic

They are roughly treated and, for example, thrown on to the backs of lorries.

MVD'S CANNOT SEE AROUND CORNERS, PARKED VEHICLES OR PLANT AND MATERIALS!

TO ENSURE EFFICIENT USE OF VA

—**USE ONLY APPROVED EQUIPMENT IN GOOD WORKING ORDER** — see page 7

—**TEST THE EQUIPMENT BEFORE SETTING UP** (and recheck it daily) — see pages 8 to 11

—**SET UP THE EQUIPMENT CORRECTLY** — see pages 12 to 16

- Use the 'VA' setting on the controller
- Align the MVD's carefully at each end of the site
- Use the correct ALL RED setting
- Use the correct MAXIMUM GREEN setting

HOW TO KNOW IF THE EQUIPMENT IS APPROVED:

Look for the 'Crown' label

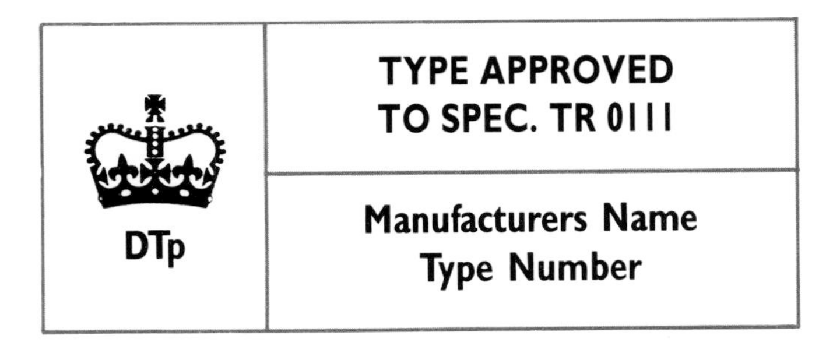

All items of equipment (controllers, signal heads and MVD's) must be fitted with a label similar to that shown above.

The current controller specification is TR 0111.

HOW TO CARRY OUT THE INIT

1. **Point signals away from the road so they cannot be seen by drivers**

2. **Set the controller to 'VA'.**

3. **Set both the 'ALL RED' controls to 5 seconds**

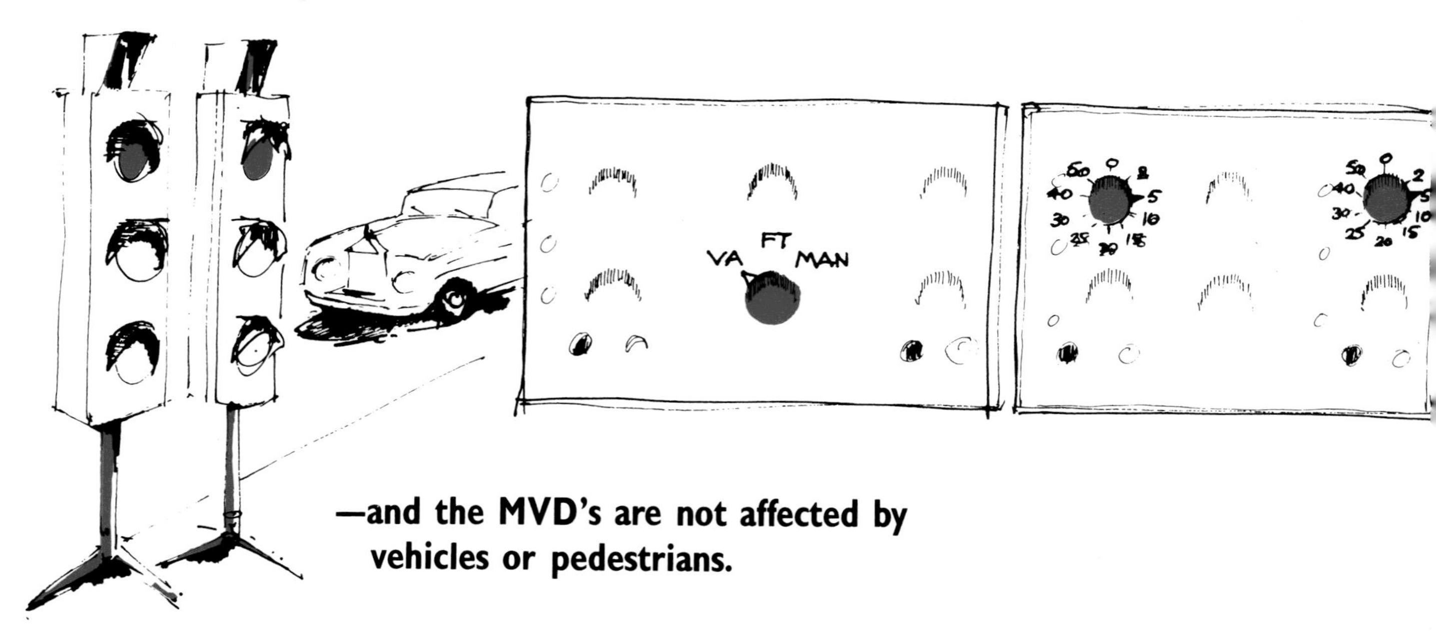

—and the MVD's are not affected by vehicles or pedestrians.

L TEST OF THE EQUIPMENT:

Set the 'MAXIMUM GREEN' controls to 15 seconds

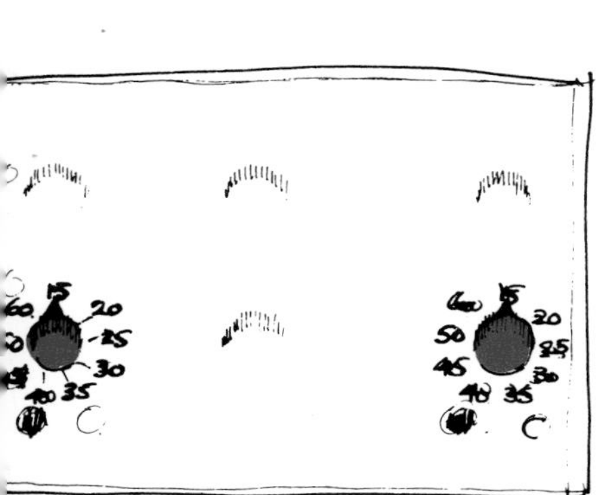

5. Connect the signal heads and power supply to the controller

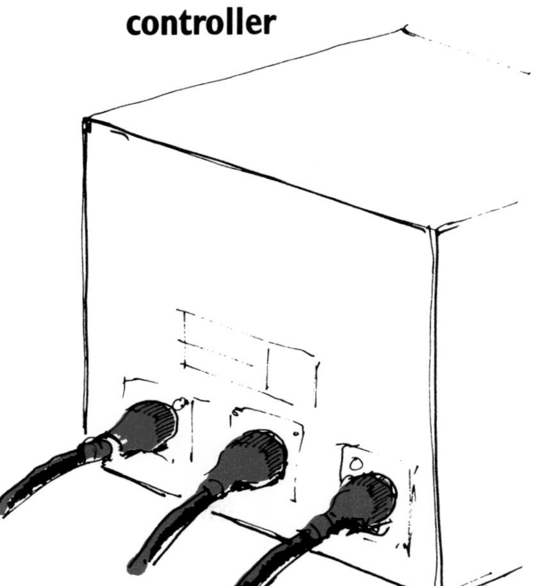

6. Switch on.

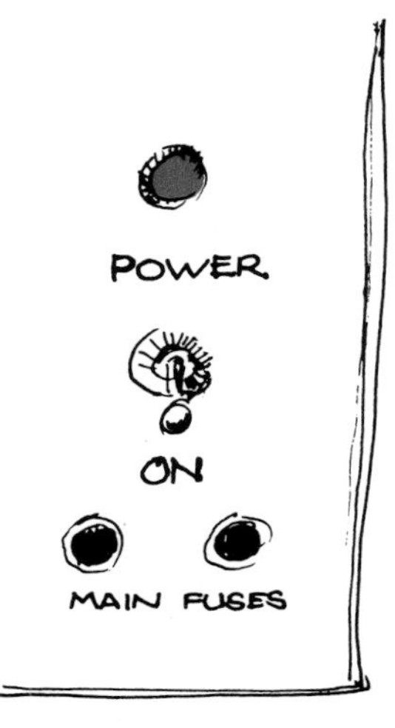

WHAT SHOULD HAPPEN:

The sequence should start with both signals showing red

They should then change in the following sequence and then rest on 'ALL RED'.

First Signal	Second Signal	shown for
RED & AMBER	RED	2 seconds
GREEN		12 seconds
AMBER		3 seconds
RED		5 seconds
	RED & AMBER	2 seconds
	GREEN	12 seconds
	AMBER	3 seconds
	RED	—

Some designs may repeat the sequence before coming to rest.

NOW CHECK THE DETECTORS:

Stand in front of one of the detectors and wave your hand smartly towards it.

This should cause the lights controlled by that detector to change. One will show green for 12 seconds and then return through amber to red.

Then check the other detector in the same manner.

IF THE SIGNALS ARE FAULTY DO NOT USE THEM—CALL OUT THE SERVICE ENGINEER.

SETTING UP THE EQUIPMENT:

Position a signal head on its tripod at each end of the site. There must not be any obstructions between the signals and vehicles up to 70 metres away.

The MVDs will generally work best if the signal heads at both ends of the site are on the nearside of the road.

If this layout is used care should be taken to allow sufficient room for traffic to pass between the end of the works and traffic waiting at the signal. The 'WAIT HERE' signs should be placed about two metres from the signal head.

If the signal head is not within the coned-off area protect it with traffic cones.

If the signals are on the near side of the road cables may have to cross the shuttle lane.

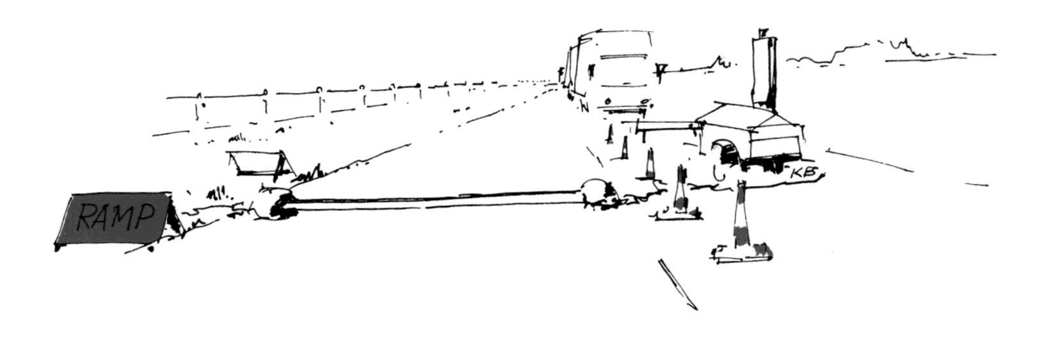

Where traffic has to pass over the cable a Cable Crossing Protector must be used.

Cable Crossing Protectors are designed to prevent damage to the cable and to permit traffic, particularly two-wheeled vehicles, to cross safely.

Drivers must be warned of the presence of the Cable Crossing Protector by means of a RAMP sign.

ADJUSTING THE RED TIMERS:

Measure the length of the works (between the two "WAIT HERE" signs). Read the appropriate red time from the scale below:

LENGTH OF WORKS (METRES)	0–50	50–100	100–150	150–200	200–250	250–300
ALL-RED TIME (SECONDS)	5	10	15	20	25	30

If the site is on a steep gradient, increase the indicated ALL-RED value for the uphill direction by 5 seconds.

If there are large numbers of slow moving vehicles which have difficulty in clearing the works before the lights have changed, increase indicated values of BOTH all-red settings by 5 seconds.

Set the 'all-red' switches to the appropriate value as determined above.

ADJUSTING THE GREEN SETTINGS:

Determine the green time for the length of works (between the "WAIT HERE" signs) from this scale:

LENGTH OF WORKS (METRES)	30–75	75–135	135–195	195–300
GREEN TIME (SECONDS)	35	40	45	50

Set both the maximum green times to the appropriate setting.

IF SUBSTANTIAL QUEUES BEGIN TO FORM AND VEHICLES TAKE MORE THAN ONE GREEN PERIOD TO GET THROUGH THE SITE THEN THESE SETTINGS ARE PROBABLY TOO SHORT AND SHOULD BE INCREASED.

BRINGING SIGNALS INTO OPERATION:

Switch controller to 'MAN' and turn Manual Control to R/R (all-red)

Watch the traffic carefully and turn one signal head to face oncoming traffic when it is safe to do so.

At the other end of the works turn that signal head to face oncoming traffic when it is safe to do so.

Switch the controller to VA when shuttle lane is clear of traffic.

NOTE:

It is important to watch the traffic at intervals and adjust, if necessary, the controller settings.

When turning the signal heads, ensure the MVD's are carefully aligned to face traffic as described on page 12.

TROUBLE-SHOOTING GUIDE

PROBLEM	POSSIBLE CAUSE	REMEDY
Very long traffic queues	1. Green setting too short 2. Faulty detectors 3. Road Capacity exceeded	1. Increase setting 2. Call service engineer 3. Discuss with highway authority urgently.
Signals do not change after one stream has stopped even though traffic is waiting	1. Faulty detectors	1. Call service engineer, work signals MANUALLY or FT until he arrives.
Green period always same length	1. Detector fault 2. Green setting too short 3. Traffic density very small	1. Call service Engineer 2. Increase green setting 3. No action required

PROBLEM	POSSIBLE CAUSE	REMEDY
Traffic stream still in shuttle lane at start of next green	1. Traffic entering shuttle lane after red has appeared 2. ALL-RED too short	1. If frequent, report facts to police 2. Increase setting
Long gap between last vehicle clearing shuttle lane and start of next green	1. ALL-RED setting too long 2. Detector fault—working Fixed time	1. Decrease setting 2. Call service engineer
Signals do not remain on red in absence of traffic	1. Detector Fault	1. Call service engineer